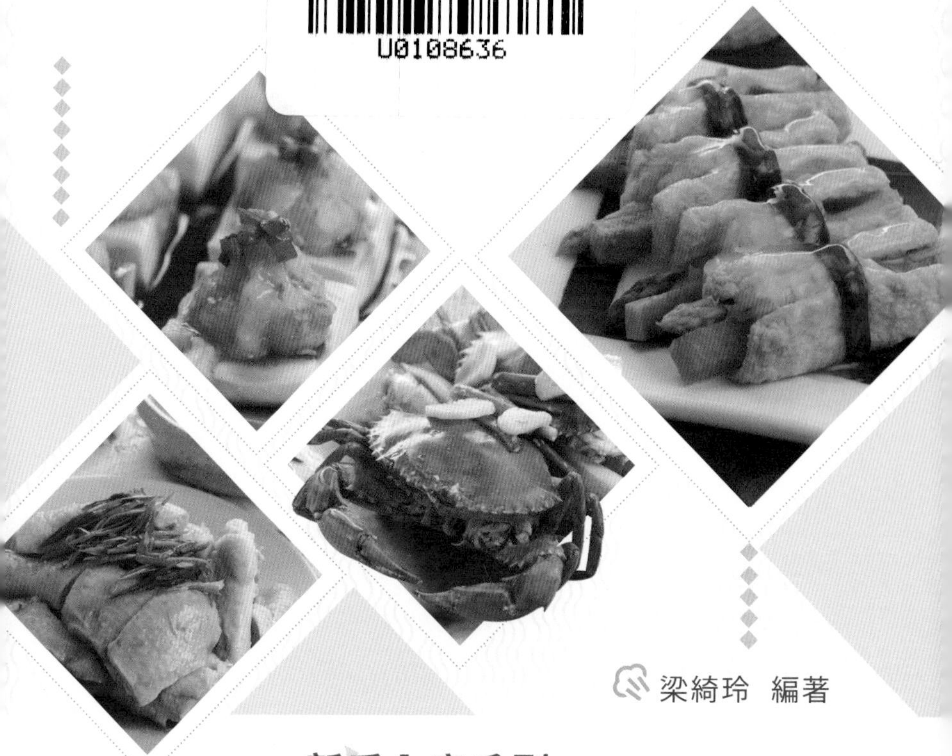

梁綺玲 編著

新手入廚系列

蒸出美味

前言

在繁忙工作後，回家吃頓豐富的晚餐，喝一碗老火湯，是人皆嚮往的樂事。無奈不少上班族平日都忙得不可開交，食得最多的是油水充足、鹽糖味精不絕的快餐，可是下班後要吃住家飯菜，也非遙不可及！

利用蒸的方法來烹煮新鮮食材，優點是可以保持食物的水分，而營養素流失也較少。既吃得到食物原味，又少油煙，是受歡迎的烹調方法。

本書以蒸餸為主題，介紹食材的名稱和選擇要點、食材的基本處理方法、蒸餸的原理和優點，以及多道簡單易學、美味家常的菜式，無論是新手入廚，抑或具有入廚經驗，都可從中取經，提高廚藝。

目錄

Buy Ingredients According to the Photos

梅頭豬肉（蒸的豬肉）
略帶肥肉。

Tenderloin pork (pork for steaming):
with some fat.

鮮雞
肉有光澤。

Chicken:
with shiny meat.

牛肉（蒸的牛肉）
瘦肉中有少許肥肉。

Beef (beef for steaming):
some fat in the lean meat.

烏頭
眼睛清晰，鱗片無脫落。

Mullet:
clear eyes, scales are not off.

鯇魚腩
要選色澤鮮明，鱗片無脫落的。

Grass carp belly:
colorful, scales are not off.

土魷
有陣陣魷魚香味。

Dried squid:
aroma of squid.

奄仔蟹
眼睛要靈活。

Roe crab (Him Zai):
lively eyes.

獅子魚
選淺黃色的，不要選黃色太鮮的。

Lion fish:
pick the slightly yellowish instead
of the brightly yellowish.

茄子
表皮光滑，重手。

Eggplant:
heavy with smooth skin.

豆腐
要選熱或暖的。

Beancurd:
pick the hot or warm.

竹笙
無雜質，無霉味。

Bamboo fungus:
without impurities and bad smell.

冬菜
淺啡色，有陣陣香味。

Spiced cabbage:
light brown and aromatic.

麵豉醬
麵豉醬可分深淺色兩種，深色麵豉醬
一般發酵期較長，味道亦較濃。

Fermented bean paste:
there are light and dark pastes.
The taste of dark fermented bean
paste is richer as the fermentation
is longer.

榨菜
內地榨菜較鹹，口感較韌。台
灣榨菜較淡，口感較硬。

Preserved mustard head:
Mainland preserved mustard
head is saltier and tough
while Taiwan preserved
mustard head is more
insipid and hard.

欖角
皮薄肉厚，肉紋幼嫩，含
油量高，味道芳香。

Preserved olives:
thin peel and thick pulp,
oily and aromatic.

鹹檸檬
完整，色澤鮮明。

Salted lemon:
intact and bright.

洗魚 Wash fish

魚肚裏的黑色膜一定要清洗乾淨，否則會有腥味。
用酒塗抹勻魚身，也可除腥味。

The black thin layer in the fish maw must be washed thoroughly, otherwise there'll be fishy smell. Smearing wine over the fish can also remove fishy smell.

處理茄子 Handle eggplant

以滾刀塊切茄子，茄子切開容易氧化變黑，可放水中略浸。

Cut eggplant into wedges. Eggplant will easily oxidize and turn black, soak in water for a while.

剁豬肉 Chop pork

先細切大剁，即先將豬肉切幼細再略剁，其次是先攪拌再搋，邊攪拌邊加水。

Strip first, that's to cut into strips first and then chop slightly. The other way is to mince the pork first, and then smash. Add water while mincing.

蒸
Steam

原理 Principle

使用高溫水蒸氣作為傳熱媒介，利用高熱將食材加熱煮熟。

Heat the ingredients with hot vapor till cooked.

優點 Advantage

保持食材的原汁原味和形態。食材的營養素流失較少。

Keep the original tastes and shapes of the food. Nutrients loss is less.

金銀蒜蒸開邊蝦

Steamed Shrimps with Minced Garlic

◯◯◯ 材料 | Ingredients

中蝦 480 克
蒜茸 3 湯匙
葱花 1 湯匙

480g medium shrimps
3 tbsps minced garlic
1 tbsp chopped spring onion

2~4 人
Serves 2~4

10 分鐘
10 minutes

⊙⊙⊙ 醃料 | Marinade

鹽 1 茶匙
胡椒粉少許

1 tsp salt
some pepper

⊙⊙⊙ 調味料 | Seasonings

蒸魚豉油 2 湯匙
油 1 湯匙

2 tbsps steamed fish soy sauce
1 tbsp oil

⊙⊙⊙ 做法 | Method

1. 蝦去殼留尾部，去腸，開邊，洗淨，瀝乾水分。下醃料拌勻略醃，將蝦鋪在蒸碟上。
2. 在鑊中放蒸架，加水至接近蒸架高度，燒滾水，將中蝦用大火隔水蒸約 5 分鐘。熄火，取出，倒去水分。
3. 另燒熱油，下一半蒜茸炸香，放在中蝦上，灑下葱花。
4. 再燒滾油，下另一半蒜茸略炒，熄火，下蒸魚豉油略煮，淋在中蝦上即成。

1. Shell shrimps and keep tails. Devein and cut into halves. Wash and drain. Add marinade and mix well. Place the shrimps on dish.
2. Place steam rack into the wok, add water to nearly the height of the rack. Boil water, steam medium shrimps over high heat for about 5 minutes. Turn off the heat and take out. Discard water.
3. Heat oil, add half of the minced garlic and deep-fry till aromatic, put it over medium shrimps and add chopped spring onion.
4. Heat oil again, add the rest of minced garlic and shallow-fry, turn off the heat. Add steam fish soy sauce and cook for a while, pour onto shrimps and serve.

入廚貴士 | Cooking Tips

- 從蝦的背部剮一刀，很容易開邊。
- It is easy to cut shrimps into halves if slightly cut at the back.

2~4 人
Serves 2~4

10 分鐘
10 minutes

豉汁蒸帶子豆腐

Steamed Scallops and Beancurd with Black Bean Sauce

材料 | Ingredients

帶子 8 隻
軟豆腐 1 磚
豆豉 1 茶匙
蒜茸 1 茶匙
葱花 1 茶匙

8 scallops
1 piece soft beancurd
1 tsp black beans
1 tsp minced garlic
1 tsp chopped spring onion

醃料 | Marinade

胡椒粉少許
some pepper

調味料 | Seasonings

蒸魚豉油1湯匙
1 tbsp steamed fish soy sauce

做法 | Method

1. 帶子洗淨，瀝乾水分，下醃料略醃。
2. 豆豉洗淨，瀝乾水分。豆腐洗淨，瀝乾水分，切成8塊。
3. 豆腐放在蒸碟上，每塊豆腐上放一粒帶子，再鋪上蒜茸和豆豉。
4. 在鑊中放蒸架，加水至接近蒸架高度，燒滾水，將帶子豆腐用大火隔水蒸約5分鐘。熄火，取出，倒去水分，灑上蔥花。
5. 另燒熱油，熄火，下蒸魚豉油略煮，淋在帶子面即成。

1. Wash scallops and drain. Marinate for a while.
2. Wash black beans and drain. Wash beancurd and drain. Cut into 8 pieces.
3. Put beancurd onto a dish, place a scallop on each beancurd, then top with minced garlic and black beans.
4. Place steam rack into wok, add water to nearly the height of the rack. Boil water, steam scallops and beancurd over high heat for about 5 minutes. Turn off the heat and take out. Discard water and add chopped spring onion.
5. Heat oil, turn off the heat. Add steamed fish soy sauce and cook for a while, pour onto scallops and serve.

入廚貼士 | Cooking Tips

- 板豆腐分軟和硬，軟豆腐宜蒸，硬豆腐宜煎。
- There are soft and hard pressed beancurds. Soft is for steam while hard is for shallow-fry.

雞油蒸奄仔蟹

Steamed Roe Crab with Chicken Fat

⊗⊗ 材料 | Ingredients

奄仔蟹 4 隻
薑片 6 片
葱段 2 棵
蒜片 2 粒
雞油 20 克
鹽適量

4 roe crabs (Him Zai)
6 ginger slices
2 sprigs sectioned spring onions
2 cloves garlic slices
20g chicken fat
some salt

2~4 人
Serves 2~4

25 分鐘
25 minutes

蘸汁 | Dipping Sauce

浙醋 1/4 杯

1/4 cup red vinegar

做法 | Method

1. 蟹劏好、洗淨,瀝乾水分,斬件,蟹鉗略拍碎。
2. 燒熱雞油,爆香蒜片備用。
3. 蟹件放在蒸碟上,加上薑片和葱段,下少許鹽,再淋上雞油和蒜片。
4. 在鑊中放蒸架,加水至接近蒸架高度,燒滾水,將蟹用大火隔水蒸約 15 分鐘,熄火,取出,與浙醋同上即成。

1. Wash and cut crabs well. Drain and cut into pieces. Crack crab claws slightly.
2. Heat chicken fat and sauté garlic slices.
3. Put crab pieces onto a dish, add ginger slices and shredded spring onion, top with some salt. Pour chicken oil and garlic slices over.
4. Place steam rack into wok, add water to nearly the height of the rack. Boil water, steam crabs over high heat for about 15 minutes. Turn off the heat and take out. Serve with red vinegar.

入廚貼士 | Cooking Tips

- 可從雞皮下取出雞膏,慢火煎出即成雞油。
- Take out the fat under chicken skin, shallow-fry it into oil.

2~4 人
Serves 2~4

25 分鐘
25 minutes

檸檬蒸烏頭

Steamed Mullet with Lemon

⬤⬤⬤ 材料 | Ingredients

烏頭 1 條（約 600 克）	1 mullet (about 600g)
鹹檸檬 1 個	1 salted lemon
薑 2 片	2 ginger slices
葱段 1 湯匙	1 tbsp sectioned spring onion
冬菇絲 1 湯匙	1 tbsp shredded dried black mushrooms
薑絲 1/2 湯匙	1/2 tbsp shredded ginger
紅辣椒絲 1 茶匙	1 tsp shredded red chili

⊙⊙ 醃料 | Marinade

鹽 1 湯匙

1 tbsp salt

⊙⊙ 調味料 | Seasonings

糖 1 湯匙	1 tbsp sugar
魚露 1 茶匙	1 tsp fish sauce
紹酒 1 茶匙	1 tsp Shaoxing wine
生粉 1 茶匙	1 tsp cornstarch
麻油少許	some sesame oil
胡椒粉少許	some pepper

⊙⊙ 做法 | Method

1. 烏頭劏好洗淨，瀝乾水分，下醃料略醃。
2. 鹹檸檬去果囊和核，切絲。
3. 將薑片和葱段鋪在蒸碟上，放上烏頭。
4. 鹹檸檬絲、薑絲、紅辣椒絲、冬菇絲與調味料拌勻，鋪在烏頭上。
5. 在鑊中放蒸架，加水至接近蒸架高度，燒滾水，將烏頭用大火隔水蒸約 12 分鐘，熄火，取出即成。

1. Wash and cut mullet well. Drain and marinate for a while.
2. Remove pulp and core of salted lemon. Shred.
3. Put shredded ginger and spring onion onto a dish, and then mullet.
4. Mix shredded salted lemon, ginger, red chili, mushroom and marinate well, and then place onto the mullet.
5. Place steam rack into wok, add water to nearly the height of the rack. Boil water, steam mullet over high heat for about 12 minutes. Turn off the heat and take out. Serve.

入廚貼士 | Cooking Tips

- 魚檔間中有游水烏頭售賣，為鹹水烏頭，無泥味，肉質鮮美，值得一試。
- From time to time fresh mullets are available at fish stalls. They are sea mullets without mud taste. The fish flesh is fresh and worth a shot.

清蒸海上鮮

Steamed Fish

材料 | Ingredients

鱸魚1條（500克）
蔥段1湯匙
薑絲1湯匙
蔥絲1湯匙
胡椒粉少許

1 bass (500g)
1 tbsp sectioned spring onion
1 tbsp shredded ginger
1 tbsp shredded spring onion
dash of pepper

2~4 人
Serves 2~4

15 分鐘
15 minutes

調味料 | Seasonings

蒸魚豉油 2 茶匙　　2 tsps steamed fish soy sauce

做法 | Method

1. 鱸魚劏好，洗淨，瀝乾水分，魚身斜剮兩刀，抹上少許胡椒粉。
2. 將一半薑絲和葱段鋪在蒸碟上，放上鱸魚。
3. 在鑊中放蒸架，加水至接近蒸架高度，燒滾水，將鱸魚用大火隔水蒸約 8 分鐘。
4. 熄火，取出鱸魚，倒去碟中蒸魚汁和棄去薑、葱。
5. 另燒熱 2 湯匙油，略爆香薑絲，熄火，下葱絲、蒸魚豉油拌匀，淋上鱸魚面即成。

1. Cut bass well, wash and drain. Slice fish slightly and sprinkle with some pepper.
2. Place half of the shredded ginger and spring onion onto the dish, and then the bass.
3. Place steam rack into the wok, add water to nearly the height of the rack. Boil water, steam bass over high heat for 8 minutes.
4. Remove from heat, take bass out. Pour out water of the steamed fish and remove ginger and spring onion.
5. Heat 2 tbsps of oil, sauté shredded ginger, remove from heat. Put in shredded spring onion and blend with steamed fish soy sauce thoroughly. Ready to serve after pouring it over the bass.

入廚貼士 | Cooking Tips

- 蒸魚要待鑊中的水燒至大滾才把魚放入；最好預先把碟也蒸熱，可以保持魚的肉質新鮮。
- Put fish into the wok only when water boils. It would be best to have the dish heated in advance to ensure the freshness of the fish.

海鮮扒蛋白

Steamed Egg White with Seafood

⟨⟨⟩⟩ 材料 | Ingredients

蝦仁 8 隻	8 shrimps
蟹柳 1 條	1 crab stick
蜆肉 20 克	20g clams
雞蛋白 3 隻	3 egg whites
清雞湯 1/2 杯	1/2 cup chicken broth
薑 2 片	2 slices ginger

⟨⟨⟩⟩ 醃料 | Marinade

胡椒粉 1 茶匙	1 tsp pepper
生粉 1/2 茶匙	1/2 tsp caltrop starch
糖 1/2 茶匙	1/2 tsp sugar
鹽 1/4 茶匙	1/4 tsp salt

◯◯◯ 調味料 | Seasonings

生抽少許　　　　Some light soy sauce

◯◯◯ 做法 | Method

1. 蜆肉、蝦仁分別洗淨，瀝乾水分，加入醃料醃 5 分鐘。蟹柳洗淨，瀝乾水分。
2. 雞蛋白放大碗中打拂，加入清雞湯拌勻，倒入深蒸碟中，蓋上錫紙。
3. 在鑊中放蒸架，加水至接近蒸架高度，燒滾水，將雞蛋白用中火隔水蒸 8–10 分鐘至熟。熄火，取出雞蛋白。
4. 另燒熱油 2 湯匙，爆香薑片，下蝦仁、蟹柳和蜆肉炒至熟，放在已蒸熟的雞蛋白上，淋上少許生抽即成。

1. Wash clams and shrimps respectively, drain. Marinate for 5 minutes. Wash and drain crab stick.
2. Beat egg white in a big bowl, mix well with chicken broth and pour into the dish. Cover with a sheet of aluminum foil.
3. Place steam rack into wok, add water to nearly the height of the rack. Boil water, steam egg white over medium heat for 8–10 minutes. Remove from heat and take egg white out.
4. Heat 2 tbsps of oil, sauté shredded ginger, stir–fry shrimps, crab stick and clams until done. Place them on the cooked egg. Pour some light soy sauce and serve.

入廚貼士 | Cooking Tips
- 蒸雞蛋時蓋上錫紙，可使雞蛋蒸熟時更加滑。
- Cover the egg with a sheet of aluminum foil when steaming can make the egg even smoother.

欖角蒸魚腩

Steamed Fish Belly with Preserved Olives

材料 | Ingredients

鯇魚腩 1 段 (400 克)	400g grass carp belly
油欖角 12 粒	12 preserved olives
葱段 1 湯匙	1 tbsp sectioned spring onion
薑 2 片	2 ginger slices
葱花 1 湯匙	1 tbsp shredded spring onion
薑絲 1/2 湯匙	1/2 tbsp shredded ginger
陳皮絲 1 茶匙	1 tsp dried citrus peel
油 1 湯匙	1 tbsp oil
蒸魚豉油 1 湯匙	1 tbsp steamed fish soy sauce
米酒適量	Some rice wine

醃料 | Marinade

鹽 1 茶匙	1 tsp salt
胡椒粉少許	some pepper

2~4 人
Serves 2~4

18 分鐘
18 minutes

26

◎◎ 做法 | Method

1. 鯇魚腩去鱗洗淨，瀝乾水分，用米酒抹勻魚身，下醃料略醃。
2. 將薑片和葱段鋪在蒸碟上，放上鯇魚腩。
3. 油欖角剁碎。將油欖角碎、薑絲、陳皮絲鋪在鯇魚腩上。
4. 在鑊中放蒸架，加水至接近蒸架高度，燒滾水，將鯇魚腩用大火隔水蒸約 12 分鐘。
5. 熄火，取出鯇魚腩，灑下葱花。
6. 另燒熱 2 湯匙油，熄火，下蒸魚豉油略煮，淋在鯇魚腩上即成。

1. Scale and wash grass carp belly, drain and smear rice wine over the fish, add some marinade.
2. Place sliced ginger and spring onion onto the dish, and then grass carp belly.
3. Shred preserved olives. Put shredded olives, ginger and dried citrus peel onto the grass carp belly.
4. Place steam rack into wok, add water to nearly the height of the rack. Boil water, steam grass carp belly for about 12 minutes over high heat.
5. Remove from heat, take out grass carp belly and add shredded spring onion.
6. Heat 2 tbsps of oil, turn off heat, add steamed fish soy sauce and cook for a while, pour on top of grass carp belly. Serve.

入廚貼士 | Cooking Tips
- 要徹底清除鯇魚腩的黑色薄膜，否則會有腥味。
- Thoroughly remove black thin film of grass carp belly, otherwise it would taste fishy.

2~4 人
Serves 2~4

15 分鐘
15 minutes

冬菜蒸龍脷柳

Steamed Basa Fillet with Spiced Cabbage

材料 | Ingredients

龍脷柳 2 條（400 克）	2 basa fillets (400g)
冬菜 1 湯匙	1 tbsp spiced cabbage
葱花 1 湯匙	1 tbsp shredded spring onion
蒸魚豉油 1 湯匙	1 tbsp steamed fish soy sauce
粉絲少許	some vermicelli
米酒適量	some rice wine
粟粉適量	some cornstarch

醃料 | Marinade

鹽 1 茶匙	1 tsp salt
胡椒粉少許	some pepper

做法 | Method

1. 龍脷柳解凍後洗淨，瀝乾水分，用米酒抹勻魚肉，下醃料略醃，塗抹少許粟粉。
2. 冬菜洗淨，瀝乾水分備用。粉絲浸軟，瀝乾水分。
3. 將粉絲鋪在蒸碟上，放上龍脷柳，冬菜鋪在龍脷柳上。
4. 在鑊中放蒸架，加水至接近蒸架高度，燒滾水，將龍脷柳用大火隔水蒸約 8 分鐘。
5. 熄火，取出龍脷柳，灑下葱花。
6. 另燒熱 2 湯匙油，熄火，下蒸魚豉油略煮，淋在龍脷柳上即成。

1. Wash basa fillets after defrosting, drain. Smear rice wine over fish fillets, marinate for a while, coat with some cornstarch.
2. Wash spiced cabbage and drain. Soak vermicelli till soft and drain.
3. Place vermicelli onto a dish, and then basa fillets, top with spiced cabbage.
4. Place steam rack into wok, add water to nearly the height of the rack. Boil water, steam basa fillets over high heat for about 8 minutes.
5. Turn off heat, take out basa fillets and add shredded spring onion.
6. Heat 2 tbsp of oil, turn off heat, add steamed fish soy sauce and cook for a while, pour on top of basa fillets. Serve.

入廚貼士 | Cooking Tips

- 龍脷柳解凍後要立即煮，解凍後不要再放回冰箱，否則肉質會變霉。
- Cook basa fillets right after defrosting. Do not put it back into the fridge after defrosting or else it would be ruined.

Steamed Lion Fish with Fermented Bean Paste

麵醬蒸獅子魚

2~4 人
Serves 2~4

10 分鐘
10 minutes

材料 | Ingredients

獅子魚 800 克	800g lion fish
葱段 1 湯匙	1 tbsp sectioned spring onion
薑 2 片	2 ginger slices
米酒適量	some rice wine

醃料 | Marinade

麵豉醬 1 湯匙	1 tbsp fermented bean paste
胡椒粉少許	some pepper

做法 | Method

1. 獅子魚劏好洗淨，瀝乾水分，用米酒抹勻魚肚，下醃料略醃。
2. 將薑切絲和葱段鋪在蒸碟上，放上獅子魚。
3. 在鑊中放蒸架，加水至接近蒸架高度，燒滾水，將獅子魚用大火隔水蒸約 5 分鐘。熄火，取出獅子魚即成。

1. Cut and wash lion fish well, drain. Smear rice wine onto fish maw, marinate for a while.
2. Place ginger shreds and sectioned spring onion onto a dish, and then lion fish.
3. Place steam rack into wok, add water to nearly the height of the rack. Boil water, steam lion fish over high heat for about 5 minutes. Turn off heat, take out lion fish and serve.

入廚貴士 | Cooking Tips

- 用米酒抹魚身或醃魚時加胡椒粉，可去除魚腥味。
- Smear rice wine on the fish or marinate fish with pepper can remove unpleasant smell.

2~4 人
Serves 2~4

20 分鐘
20 minutes

蒸釀蟹蓋

Steamed Stuffed Crab

◯◯◯ 材料 | Ingredients

奄仔蟹 4 隻
絞豬肉 300 克
雞蛋 1 隻

4 roe crabs (Him Zai)
300g minced pork
1 egg

醃料 | Marinade

生抽 2 茶匙	2 tsps soy sauce
生粉 1 茶匙	1 tsp caltrop starch
胡椒粉少許	some pepper

做法 | Method

1. 奄仔蟹拆下蟹蓋，劏好，刷洗淨，瀝乾水分備用。

2. 絞豬肉加醃料略醃 15 分鐘，加入雞蛋拌勻，釀入蟹蓋中。將蟹蓋放在蒸碟上。

3. 在鑊中放蒸架，加水至接近蒸架高度，燒滾水，將蟹蓋用大火隔水蒸約 10 分鐘，熄火，取出即成。

1. Remove shells from roe crabs, cut well, brush, wash and drain.
2. Marinate minced pork for 15 minutes, add egg and mix, then stuff it into crab shells. Place crab shells onto a dish.
3. Place steam rack into wok, add water to nearly the height of the rack. Boil water, steam stuffed crabs over high heat for about 10 minutes. Turn off heat and take out. Serve.

入廚貼士 | Cooking Tips

- 購買蟹回家後，若不是即時煮食，要放入雪櫃，否則容易變壞。
- Put the crab into the fridge if it's not cooked immediately, otherwise it will be ruined easily.

蝦醬蒸魷魚筒

Steamed Squid with Shrimp Paste

2~4 人
Serves 2~4

10 分鐘
10 minutes

材料 | Ingredients

魷魚筒 350 克
葱花 1/2 湯匙

350g squid
1/2 tbsp chopped spring onion

醃料 | Marinade

蝦醬 1 湯匙
糖 1 茶匙
胡椒粉少許

1 tbsp shrimp paste
1 tsp sugar
some pepper

做法 | Method

1. 魷魚筒洗淨，瀝乾水分。
2. 蝦醬下糖拌勻。魷魚筒加蝦醬、胡椒粉略醃 15 分鐘。
3. 在鑊中放蒸架，加水至接近蒸架高度，燒滾水，將魷魚筒用大火隔水蒸約 4 分鐘，熄火，取出，加葱花即成。

1. Wash and drain squid.
2. Add sugar in shrimp paste and mix well. Slightly marinate squid with shrimp paste and pepper for 15 minutes.
3. Place steam rack into wok, add water to nearly the height of the rack. Boil water, steam squid over high heat for about 4 minutes. Turn off heat and take out. Add spring onion and serve.

入廚貼士 | Cooking Tips

- 蝦醬若不加糖，味道會太鹹。
- Shrimp paste will be too salty without adding sugar.

百花蒸釀豆腐

Steamed Stuffed Beancurd with Shrimp

⦿ 材料 | Ingredients

蝦 100 克	100g shrimps
軟豆腐 1 磚	1 piece soft beancurd
葱花 1 茶匙	1 tsp chopped spring onion
粟粉 1 茶匙	1 tsp cornstarch
鹽水適量	some salt water

⦿ 醃料 | Marinade

鹽 1 茶匙	1 tsp salt
胡椒粉少許	some pepper

⦿ 調味料 | Seasonings

蒸魚豉油 1 湯匙	1 tbsp steamed fish soy sauce

⚪⚪ 做法 | Method

1. 蝦去殼去腸，洗淨，瀝乾水分。用鹽水略浸，再瀝乾水分。

2. 蝦放在砧板上用刀背略拍，再剁，放在大碗中，下醃料拌勻，順一個方向用力攪拌至起膠。

3. 豆腐洗淨，瀝乾水分，切成 8 塊，放在蒸碟上，豆腐面上掃上粟粉，釀入蝦膠。

4. 在鑊中放蒸架，加水至接近蒸架高度，燒滾水，將釀豆腐用大火隔水蒸約 8 分鐘，熄火，取出，倒去碟中水分，加葱花。

5. 燒滾油，熄火，下蒸魚豉油略煮，淋在蝦膠上即成。

1. Shell and devein shrimps. Wash and drain. Soak slightly with salt water, and then drain again.

2. Put shrimps on chopping board and slightly pat with the back of knife, chop then. Put it into a big bowl, mix with marinade. Stir in one direction till elastic.

3. Wash beancurd and drain, cut into 8 pieces and put them on dish. Coat the beancurd surfaces with cornstarch, stuff with shrimp paste.

4. Place steam rack into wok, add water to nearly the height of the rack. Boil water, steam stuffed beancurd over high heat for about 8 minutes. Turn off heat and take out. Pour out water in dish and add spring onion.

5. Heat oil, turn off heat. Add steamed fish soy sauce and cook slightly, pour over shrimp paste and serve.

入廚貼士 | Cooking Tips

- 蝦蒸煮前用鹽水略浸，可保持肉質爽口。
- Slightly soak shrimps in salt water before steaming to ensure the texture is elastic.

鮮蝦油豆腐

Fried Beancurd with Shrimp

材料 | Ingredients

鮮蝦 6 隻	6 shrimps
火腿扒 1 片	1 ham steak
蘆筍 6 條	6 asparagus
日本油揚 6 個	6 yaki aburaage
韭菜 6 條	6 leeks

芡汁料 | Thickening

清湯 1/4 杯	1/4 cup chicken broth
粟粉 1 湯匙	1 tbsp corn starch
糖 1/4 茶匙	1/4 tsp sugar
鹽 1/4 茶匙	1/4 tsp salt

2~4 人
Serves 2~4

15 分鐘
15 minutes

做法 | Method

1. 鮮蝦去殼去腸，洗淨，瀝乾水分。火腿扒洗淨，瀝乾水分，切粗條。蘆筍洗淨，切成火腿條般長度。

2. 韭菜和油揚分別洗淨，瀝乾水分。鑊中燒滾水，加入汆水，取出，壓出水分。

3. 把1條蘆筍、1條火腿和1隻蝦放在油揚上，用韭菜綁好，放蒸碟上。

4. 燒滾水，將油揚卷用大火隔水蒸約5分鐘至熟，熄火，取出，倒去水分。

5. 將芡汁料放碗中拌勻。燒熱鑊，下芡汁料，拌勻，煮成玻璃芡，淋在油揚上即成。

1. Shell and devein shrimps. Wash and drain. Wash and drain ham, strip. Wash asparagus, cut according to the length of the ham.

2. Wash leeks and yaki aburaage and drain. Boil water in wok, cook leeks and yaki aburaage respectively till soft. Take out and dry.

3. Put 1 asparagus, 1 ham and 1 shrimp on yaki aburaage, tie up with leeks and put onto a dish.

4. Boil water, steam yaki aburaage rolls over high heat for about 5 minutes till cooked. Remove from heat and take out, pour out the water.

5. Mix thickening in the bowl. Heat wok and pour in thickening, mix well and cook till transparent, pour over yaki aburaage and serve.

入廚貴士 | Cooking Tips

- 可因應個人口味而用其他配料代替火腿、蘆筍。
- Ham and asparagus can be substituted by other ingredients according to personal tastes.

4~6 人
Serves 4~6

15 分鐘
15 minutes

Steamed Crab

清蒸花蟹

⊙⊙ 材料 | Ingredients

花蟹1隻

1 crab

⊙⊙ 醃料 | Marinade

薑茸1茶匙

香醋2湯匙

1 tsp minced ginger
2 tbsps red vinegar

⊙⊙ 調味料 | Seasonings

花椒數粒

Some Sichuan peppercorn

⊙⊙ 做法 | Method

1. 把薑茸倒入香醋內拌勻成薑醋汁。
2. 蟹洗淨，連捆紮草繩，翻轉放入碟中，加花椒，以猛火蒸8分鐘（可熱食或待冷），蘸薑醋汁即可。

1. Mix minced ginger with red vinegar as dipping sauce.
2. Wash crab, put onto a plate upside down with straw rope, top with Sichuan peppercorn. Boil water, steam over high heat for about 8 minutes till cooked. Serve hot or cold with dipping sauce.

蒜茸蒸茄子

Steamed Eggplant with Minced Garlic

2~4 人
Serves 2~4

10 分鐘
10 minutes

材料 | Ingredients

茄子 1 條
蒜茸 1 1/2 湯匙
葱花 1 湯匙

1 eggplant
1 1/2 tbsps minced garlic
1 tbsp chopped spring onion

芡汁料 | Thickening

生抽 2 茶匙
麻油 1 茶匙

2 tsps light soy sauce
1 tsp sesame oil

做法 | Method

1. 茄子洗淨，瀝乾水分，開邊。
2. 在鑊中放蒸架，加水至接近蒸架高度，燒滾水，將茄子用大火隔水蒸約 6 分鐘，取出。
3. 灑下葱花，下麻油。
4. 燒熱油鑊，下蒜茸爆香，熄火，加生抽拌勻，淋在茄子上即成。

1. Wash and drain eggplant, cut into halves.
2. Place steam rack into wok, add water to nearly the height of the rack. Boil water, steam eggplant over high heat for about 6 minutes. Take out.
3. Spread chopped spring onion and add sesame oil.
4. Heat oil in wok, sauté minced garlic till aromatic, turn off heat. Add light soy sauce and mix, pour on eggplant and serve.

入廚貼士 | Cooking Tips

- 茄子切開後，蒸煮前要放鹽水中浸泡，可保持色澤鮮艷。
- After cutting eggplant, soak in salt water before steaming to keep it colorful.

本菇竹笙卷

Steamed Bamboo Fungus Roll with Shimeji Mushroom

材料 | Ingredients

本菇 200 克	200g shimeji mushrooms
竹笙 8 條	8 bamboo fungus
蟹柳 3 條	3 crab sticks
甘筍絲 2 湯匙	2 tbsps shredded carrot
粉絲少許	some vermicelli

醃料 | Marinade

鹽 1/2 茶匙	1/2 tsp salt
糖 1/4 茶匙	1/4 tsp sugar
麻油少許	some sesame oil

芡汁料 | Thickening

清雞湯 1/2 杯	1/2 cup chicken broth
生粉 2 茶匙	2 tsps caltrop starch
鹽 1/2 茶匙	1/2 tsp salt
麻油少許	some sesame oil
胡椒粉少許	some pepper

做法 | Method

1. 竹笙浸軟洗淨，瀝乾水分，剪去頭尾，剪成塊狀。
2. 本菇去蒂，洗淨，瀝乾水分。
3. 粉絲浸軟，瀝乾水分，切段，加入蟹柳絲、甘筍絲和醃料拌勻。
4. 燒熱鑊，加入清雞湯煮滾，加入竹笙煨片刻，盛起。
5. 竹笙塊鋪在碟上，將粉絲、蟹柳絲和甘筍絲放在竹笙上，捲成筒狀，可用牙籤固定。
6. 在鑊中放蒸架，加水至接近蒸架高度，燒滾水，將竹笙卷用大火隔水蒸約 4 分鐘，取出，倒出汁液。
7. 燒熱油鑊，下本菇略炒，加入芡汁料拌勻煮滾，淋在竹笙上即成。

1. Wash and drain bamboo fungus, cut both ends and cut into pieces.
2. Discard stalks of shimeji mushrooms. Wash and drain.
3. Soak vermicelli till soft and drain. Cut into sections. Add shredded crab stick and carrot, mix well.
4. Heat wok, boil chicken broth, add bamboo fungus and cook for a while, take out.
5. Arrange bamboo fungus onto a dish, place vermicelli, shredded crab stick and carrot onto bamboo fungus. Roll up and fix tightly with toothpicks.
6. Place steam rack into wok, add water to nearly the height of the rack. Boil water, steam bamboo fungus over high heat for about 4 minutes. Take out and pour out sauce.
7. Heat oil in wok, slightly stir-fry shimeji mushroom, add thickening, mix and cook, pour onto bamboo fungus and serve.

入廚貼士 | Cooking Tips

- 竹笙浸軟後，要剪去尾部。
- Cut the end of bamboo fungus after soaking.

三寶釀竹笙

材料 | Ingredients

竹笙 10 條	10 bamboo fungus
火腿 1 片	1 slice ham
蘆筍 25 克	25g asparagus
冬菇 3 朵	3 dried black mushrooms

芡汁料 | Thickening

清雞湯 1/4 杯	1/4 cup chicken broth
蠔油 2 茶匙	2 tsps oyster sauce
糖 1 茶匙	1 tsp sugar

4-6 人
Serves 4~6

15 分鐘
15 minutes

⭕⭕ 做法 | Method

1. 竹笙浸軟洗淨，瀝乾水分，剪去頭尾。
2. 冬菇浸軟，瀝乾水分，去蒂，切絲。
3. 火腿、蘆筍分別洗淨，瀝乾水分。火腿切條，蘆筍切段。
4. 將冬菇、火腿、蘆筍各一釀入竹笙內，排在蒸碟上，芡汁料放碗中拌勻，淋在釀竹笙上。
5. 在鑊中放蒸架，加水至接近蒸架高度，燒滾水，將釀竹笙用大火隔水蒸約 8 分鐘即成。

1. Wash and drain bamboo fungus, cut both ends.
2. Soak dried black mushrooms and drain. Discard stalks and shred.
3. Wash ham and asparagus respectively and drain. Strip ham and cut asparagus into sections.
4. Stuff dried black mushrooms, ham and asparagus into bamboo fungus, arrange onto a dish. Mix thickening in bowl and pour onto bamboo fungus.
5. Place steam rack into wok, add water to nearly the height of rack. Boil water, steam stuffed bamboo fungus over high heat for about 8 minutes and serve.

入廚貼士 | Cooking Tips
- 所有材料的長度要均一，較美觀。
- All ingredients should be of the same length.

4-6 人
Serves 4~6

15 分鐘
15 minutes

金銀蛋蒸豆花

Steamed Preserved Eggs with Beancurd

材料 | Ingredients

軟豆腐 1 磚
雞蛋 1 隻
鹹蛋 1 隻
皮蛋 1 隻
葱花 1 茶匙

1 piece soft beancurd
1 egg
1 salted egg
1 preserved egg
1 tsp chopped spring onion

調味料 | Seasonings

生抽 2 茶匙	2 tsps light soy sauce
鹽 1 茶匙	1 tsp salt
生粉 1 茶匙	1 tsp caltrop starch
油 1 茶匙	1 tsp oil

做法 | Method

1. 鹹蛋、皮蛋洗淨。雞蛋打勻。
2. 燒熱一鍋水，加入鹹蛋、皮蛋焓熟，取出，浸在凍水中冷卻，除殼，切粒。
3. 豆腐洗淨，瀝乾水分，攪碎，加調味料、雞蛋、鹹蛋和皮蛋粒拌勻，平鋪在蒸碟上。
4. 在鑊中放蒸架，加水至接近蒸架高度，燒滾水，將豆腐用大火隔水蒸約 8 分鐘，灑上葱花。
5. 燒熱油，熄火，下生抽略煮，淋在豆腐上即成。

1. Wash salted egg, preserved egg. Mix the egg well.
2. Heat a pot of water, boil salted egg and preserved egg and then take out. Soak and cool in cold water. Shell and cut into dice.
3. Wash beancurd and drain. Mince and add seasonings, egg, salted egg and preserved egg to blend well, put onto a dish flat.
4. Place steam rack into wok, add water to nearly the height of rack. Boil water, steam beancurd over high heat for about 8 minutes. Sprinkle with chopped spring onion.
5. Heat oil and remove from heat. Add light soy sauce and slightly cook, pour on beancurd and serve.

入廚貼士 | Cooking Tips

- 雞蛋焓熟後，立即浸在凍水中冷卻，較易除殼。
- It will be easier to shell egg if put egg into cold water immediately after boiling.

鹹豬肉蒸冬瓜夾

Steamed Salted Pork in Winter Melon Sandwich

冬瓜 400 克
鹹豬肉 200 克
清雞湯 1 杯

400g winter melon
200g salted pork
1 cup chicken broth

2~4 人
Serves 2~4

20 分鐘
20 minutes

50

⊙⊙ 芡汁料 | Thickening

蠔油 1 湯匙	1 tbsp oyster sauce
粟粉 2 茶匙	2 tsps cornstarch
糖 1/2 茶匙	1/2 tsp sugar
水 3 湯匙	3 tbsps water

⊙⊙ 做法 | Method

1. 冬瓜洗淨，瀝乾水分，去皮，切成長方塊，在中央直切一刀。
2. 鹹豬肉切片，釀入冬瓜夾中，排上深蒸碟上，倒下清雞湯。
3. 在鑊中放蒸架，加水至接近蒸架高度，燒滾水，將冬瓜夾用大火隔水蒸約 15 分鐘，熄火，倒出汁液。
4. 芡汁料放碗中拌勻，加入汁液拌勻。
5. 另燒熱油鑊，下芡汁料煮至濃，淋上冬瓜夾上即成。

1. Wash winter melon and drain. Peel and cut into pieces, then cut in the middle but do not cut through.
2. Slice salted pork, stuff into the middle of winter melon sandwich, place onto a dish in order and pour in chicken broth.
3. Place steam rack into wok, add water to nearly the height of the rack. Boil water, steam winter melon sandwich over high heat for about 15 minutes. Turn off heat, pour out the sauce.
4. Mix thickening into a bowl, add sauce and mix.
5. Heat oil in wok, add thickening and cook till it thickens. Pour onto winter melon sandwich and serve.

入廚貼士 | Cooking Tips

- 冬瓜中央所切的一刀，不要切到底，否則不能夾實鹹豬肉。
- Don't cut to the bottom of winter melon, otherwise it can't hold the salted pork.

2~4 人
Serves 2~4

15 分鐘
15 minutes

蝦米蒸水蛋

Steamed Egg with Dried Shrimp

材料 | Ingredients

蝦米 1 湯匙
雞蛋 3 隻
1 tbsp dried shrimps
3 eggs

調味料 | Seasonings

生抽 2 茶匙
2 tsps light soy sauce

做法 | Method

1. 蝦米浸軟，洗淨，瀝乾水分。
2. 雞蛋放大碗中，加相同份量水拌勻，下調味料拌勻，倒進深蒸碟中，蓋上錫紙。
3. 在鑊中放蒸架，加水至接近蒸架高度，燒滾水，將雞蛋用中火隔水蒸 8 分鐘至熟即成。

1. Soak dried shrimps till soft. Wash and drain.
2. Put egg in big bowl, add same amount of water, mix with seasonings. Pour into a deep dish, cover with aluminum foil.
3. Place steam rack into wok, add water to nearly the height of the rack. Boil water, steam egg over medium heat about 8 minutes till cooked. Serve.

入廚貼士 | Cooking Tips

* 用牙籤刺入水蛋，可測試是否已熟。
* Poke a toothpick into steamed egg to check if it's cooked.

海鮮茶碗蒸

Steamed Cup Custard with Seafood

材料 | Ingredients

蟹柳 1 條
蝦仁 4 隻
雞蛋 2 隻
清雞湯 3/4 杯

1 crab stick
4 shrimps
2 eggs
3/4 cup chicken broth

2 人
Serves 2

15 分鐘
15 minutes

調味料 | Seasonings

生抽 1 茶匙

1 tsp light soy sauce

做法 | Method

1. 蟹柳洗淨，瀝乾水分，撕成絲。
2. 蝦仁洗淨，瀝乾水分。將蝦仁用大火隔水蒸約 5 分鐘至熟，去殼。
3. 雞蛋在大碗中打勻，加入清雞湯拌勻，倒入兩個碗中，蓋上錫紙。
4. 在鑊中放蒸架，加水至接近蒸架高度，燒滾水，將茶碗蒸用大火隔水蒸約 6 分鐘至八成熟，取出。
5. 將蝦仁、蟹柳絲放在雞蛋面，再蒸 2 分鐘至熟，淋上生抽即

1. Wash and drain crab stick, tear into strips.
2. Wash and drain shrimps. Steam shrimps over high heat for about 5 minutes till cooked, shell.
3. Mix eggs in a big bowl, blend with chicken broth, pour into two cups, cover with aluminum foil.
4. Place steam rack into wok, add water to nearly the height of the rack. Boil water, steam cup custards over high heat for about 6 minutes till 80% cooked. Take out.
5. Place shrimps and crab stick strips on egg surface. Steam for another 2 minutes till cooked, pour light soy sauce and serve.

入廚貼士 | Cooking Tips

- 雞蛋拌勻後，先待 5 分鐘再蒸，讓雞蛋中的空氣排出，可使雞蛋較幼滑。
- After mixing the egg, wait for 5 minutes before steaming. The egg will be smoother as bubbles in egg are gone.

釀藕片

Steamed Stuffed Lotus Root Slices

材料 | Ingredients

蓮藕 1 節
免治豬肉 80 克
蔥 1 棵

1 lotus root section
80g minced pork
1 sprig spring onion

醃料 | Marinade

鹽 1 茶匙
生抽 1/2 茶匙
糖 1/2 茶匙
生粉 1/2 茶匙
油 1 茶匙

1 tsp salt
1/2 tsp light soy sauce
1/2 tsp sugar
1/2 tsp caltrop starch
1 tsp oil

芡汁料 | Thickening

蠔油 2 湯匙
生粉 2 茶匙
清水 3 湯匙
麻油少許

2 tbsps oyster sauce
2 tsps caltrop starch
3 tbsps water
some sesame oil

做法 | Method

1. 蓮藕洗淨，切齊頭尾，放入滾水中煮 8 分鐘，取出瀝乾水份，待涼。
2. 免治豬肉下醃料醃 10 分鐘，然後將肉碎釀入蓮藕中。
3. 蓮藕隔水蒸 10 分鐘，涼凍後切片。
4. 燒熱 1 湯匙油，爆香葱，下芡汁，煮至濃稠。淋在藕片上。

1. Cut both ends of lotus root, cook in boiling water for 8 minutes, take out and drain. Wait till cool.
2. Marinate minced pork for 10 minutes, and then stuff into lotus root.
3. Steam lotus root for 10 minutes, cut after cooling down.
4. Heat 1 tbsp of oil, sauté spring onion till aromatic. Add thickening and cook till it thickens. Pour onto lotus root slices.

入廚貼士 | Cooking Tips

- 釀肉時可利用筷子將肉碎塞進蓮藕縫中。
- Stuff minced meat into holes of lotus root with chopsticks.

龍穿鳳翼

Stuffed Chicken Wings

⊙⊙ 材料 | Ingredients

雞中翼 10 隻	10 chicken wings (middle part)
火腿扒 1/2 塊	1/2 ham steak
西芹 1 條	1 celery
甘筍 1 小條	1 small carrot

⊙⊙ 醃料 | Marinade

生抽 2 茶匙	2 tsps light soy sauce
酒 1 茶匙	1 tsp wine
鹽 1 茶匙	1 tsp salt
老抽 1 茶匙	1 tsp dark soy sauce
薑汁 1 茶匙	1 tsp ginger juice
糖 1/2 茶匙	1/2 tsp sugar
胡椒粉少許	some pepper

2~4 人
Serves 2~4

40 分鐘
40 minutes

做法 | Method

1. 西芹、甘筍、火腿扒分別洗淨，瀝乾水分，切成條狀。

2. 雞翼洗淨，瀝乾水分（如用雪藏雞翼需要預先解凍），切去頭尾部分的筋，去骨。

3. 雞翼加醃料醃 15 分鐘。

4. 將西芹、甘筍、火腿條分別釀入雞翼內，排在蒸碟上。

5. 在鑊中放蒸架，加水至接近蒸架高度，燒滾水，將雞翼用大火隔水蒸約 13 分鐘至熟即成。

1. Wash celery, carrot, ham steak respectively, drain and strip.
2. Wash chicken wings and drain (defrost first if frozen), cut the vein at both ends, remove bones.
3. Marinate chicken wings for 15 minutes.
4. Stuff celery, carrot and ham sticks into chicken wings, put on a dish in order.
5. Place steam rack into wok, add water to nearly the height of the rack. Boil water, steam chicken wings over high heat about 13 minutes till cooked. Serve.

入廚貼士 | Cooking Tips

- 於雞翼頭尾剔開雞翼的筋，沿雞骨向下剔，再將雞肉連皮向下推，便很容易將雞翼去骨。
- Cut the vein at both ends of chicken wing. Cut along chicken bone, push chicken meat with skin downwards, it's easy to remove bones from the chicken wing.

話梅蒸雞翼

Steamed Chicken Wing with Plum

材料 | Ingredients

雞中翼 500 克	500g chicken wings
甜話梅 5 粒	5 sweet preserved plums
金針 40 克	40g dried lily flowers
雲耳 40 克	40g black fungus
薑 2 片	2 ginger slices
葱段 1 棵量	1 sprig spring onion (sectioned)

醃料 | Marinade

薑汁 1 茶匙	1 tsp ginger juice
酒 1 茶匙	1 tsp wine
粟粉 1/2 茶匙	1/2 tsp cornstarch
胡椒粉少許	some pepper

調味料 | Seasonings

清雞湯 1/2 杯	1/2 cup chicken broth
蠔油 1 湯匙	1 tbsp oyster oil
麻油少許	some sesame oil

做法 | Method

1. 雲耳、金針分別浸透，洗淨，瀝乾水分，去蒂。雲耳切小朵。鑊中燒滾水，分別加入雲耳、金針汆水，取出過冷河，壓出水分。
2. 雞翼洗淨，瀝乾水分（如用雪藏雞翼需要預先解凍），加入醃料略醃。
3. 話梅起肉切碎。
4. 雞翼排在蒸碟上，鋪上話梅肉、金針、雲耳，下調味料拌勻。
5. 在鑊中放蒸架，加水至接近蒸架高度，燒滾水，將雞翼用大火隔水蒸約 8 分鐘至熟即成。

1. Soak thoroughly black fungus, dried lily flower. Wash and drain, discard stalks. Cut black fungus into small pieces. Boil water in a wok, blanch black fungus, dried lily flowers respectively, then take out and rinse in cold water, squeeze out water.
2. Wash and drain chicken wings (defrost first if frozen), marinate for a while.
3. Cut preserved plums into pieces, discard cores.
4. Place chicken wings onto a dish, put plums, dried lily flowers, black fugnus on top. Add seasonings and mix.
5. Place steam rack into wok, add water to nearly the height of the rack. Boil water, steam chicken wings over high heat for about 8 minutes till cooked. Serve.

入廚貼士 | Cooking Tips

- 雲耳、金針汆水，可去除霉味。
- Blanch black fungus and dried lily flowers to remove undesired smell.

酸梅蒸鴨片

Steamed Duck with Plum

材料 | Ingredients

米鴨1隻
梅子3粒
薑4片
蔥段2棵
蔥花1湯匙
紅辣椒圈1茶匙
蒜茸1茶匙

1 duck
3 sour plums
4 ginger slices
2 sprigs spring onion
1 tbsp chopped spring onion
1 tsp red chili rings
1 tsp minced garlic

4~6 人
Serves 4~6

90 分鐘
90 minutes

醃料 | Marinade

老抽 2 湯匙	2 tbsps dark soy sauce
生抽 1 湯匙	1 tbsp light soy sauce

調味料 | Seasonings

糖 2 茶匙	2 tsps sugar

做法 | Method

1. 米鴨洗淨，瀝乾水分，用醃料醃 30 分鐘，將葱段和薑片放入鴨腔內。
2. 梅子去核，放碗中用叉搓成茸。
3. 在鑊中放蒸架，加水至接近蒸架高度，燒滾水，將米鴨用大火隔水蒸約 30–40 分鐘至熟，稍涼後起肉切片。
4. 燒熱油鑊，爆香紅辣椒圈和蒜茸，加入梅子和調味料，煮滾後淋在鴨肉上，灑上葱花即成。

1. Wash and drain duck. Marinate for 30 minutes, put spring onions and ginger slices into duck.
2. Discard cores of sour plums, mince with a fork in bowl.
3. Place steam rack into wok, add water to nearly the height of the rack. Boil water, steam duck over high heat for about 30–40 minutes till cooked. Cut into pieces after it cools down a little bit.
4. Heat oil in wok, sauté red chili rings and minced garlic, add sour plums and seasonings. Pour over duck meat after cooked. Dash shredded spring onion and serve.

入廚貼士 | Cooking Tips

- 除去鴨的尾部可以減少羶味。
- To remove the tail of duck can reduce the unpleasant smell.

4~6 人
Serves 4~6

18 分鐘
18 minutes

鮮雞雲腿冬菇件

Steamed Chicken and Ham with Mushroom

材料 | Ingredients

鮮雞髀 2 隻	2 fresh chicken legs
冬菇 3 朵	3 dried black mushrooms
雲腿 10 片	10 Yunnan ham slices

醃料 | Marinade

鹽 1 湯匙	1 tbsps salt
生抽 2 茶匙	2 tsps light soy sauce
生粉 1 茶匙	1 tsp caltrop starch
米酒 1 茶匙	1 tsp rice wine
薑汁 1 茶匙	1 tsp ginger juice
糖 1/2 茶匙	1/2 tsp sugar

⊙⊙⊙ 調味料 | Seasonings

蠔油 1 茶匙	1 tsp oyster oil
生抽 1 茶匙	1 tsp light soy sauce
糖 1/2 茶匙	1/2 tsp sugar
生粉 1/2 茶匙	1/2 tsp caltrop starch

⊙⊙⊙ 做法 | Method

1. 雞髀洗淨,用醃料醃 5 分鐘,斬件。
2. 冬菇浸軟,去蒂,洗淨。雲腿洗淨,瀝乾水分。
3. 將雞件、雲腿、冬菇相間地按序排在蒸碟上,將調味料在碗中拌匀,淋在雞件、雲腿、冬菇上。
4. 在鑊中放蒸架,加水至接近蒸架高度,燒滾水,將雞件用大火隔水蒸約 12 分鐘即成。

1. Wash chicken legs, marinate for 5 minutes. Chop into pieces.
2. Soak dried black mushrooms till soft, remove stalks and wash. Wash ham and drain.
3. Put chicken, ham and mushrooms alternately onto a dish, mix seasonings in bowl. Pour it over chicken, ham and mushrooms.
4. Place steam rack into wok, add water to nearly the height of the rack. Boil water, steam chicken over high heat for about 12 minutes. Serve.

入廚貼士 | Cooking Tips

- 冬菇用生粉搓洗淨,可使色澤更鮮明。
- Rub dried black mushrooms with caltrop starch and rinse and it will look brighter.

家常蒸雞

Homemade Steamed Chicken

光雞 1 隻
乾蔥頭 3 粒
蔥絲 1 杯
薑絲 1/4 杯
紅辣椒絲 1/2 湯匙
鹽適量

1 chicken
3 shallots
1 cup shredded spring onion
1/4 cup shredded ginger
1/2 tbsp shredded red chili
some salt

15 分鐘
15 minutes

4~6 人
Serves 4~6

調味料 | Seasonings

麻油 1 湯匙　　　1 tbsp sesame oil
生抽 2 茶匙　　　2 tsps soy sauce
沙薑粉 1/2 茶匙　1/2 tsp mush ginger power

做法 | Method

1. 光雞洗淨，瀝乾水分，用鹽抹勻雞身略醃。
2. 在鑊中放蒸架，加水至接近蒸架高度，燒滾水，將雞用大火隔水蒸約 20 分鐘至熟，稍攤涼後斬件上碟。
3. 薑絲、葱絲和紅辣椒絲放碗中拌勻。
4. 燒熱油鑊，下乾葱頭爆香，棄去乾葱頭，將滾油潷入混合薑葱料中，加調味料拌勻，淋在雞上即成。

1. Wash chicken and drain. Smear salt on chicken and marinate for a while.
2. Place steam rack into wok, add water to nearly the height of the rack. Boil water, steam chicken over high heat for 20 minutes. Cool down for a while. Chop and put onto a dish.
3. Put shredded ginger, spring onion and red chili in bowl and mix.
4. Heat oil in wok, sauté shallots till aromatic and discard. Add hot oil to ginger and spring onion, add marinade and mix well. Pour over chicken and serve.

入廚貼士 | Cooking Tips
- 用長竹籤插入雞髀最厚部位，能順利插入表示雞已熟。
- Insert a bamboo stick into the thickest part of the chicken leg, it's done if it can get through.

古法蒸雞翼

Steamed Chicken Wings in Traditional Style

⊙⊙ 材料 | Ingredients

雞翼 300 克	300g chicken wings
雲耳 10 克	10g black fungus
金針 10 克	10g dried lily flowers
冬菇 4 朵	4 dried black mushrooms
杞子 1 湯匙	1 tbsp wolfberries
米酒 2 茶匙	2 tsps rice wine

⊙⊙ 醃料 | Marinade

生抽 1 湯匙	1 tbsp soy sauce
薑汁 2 茶匙	2 tsps ginger juice
酒 2 茶匙	2 tsps wine
魚露 1 茶匙	1 tsp fish sauce
糖 1/2 茶匙	1/2 tsp sugar
麻油少許	some sesame oil

調味料 | Seasonings

上湯 1/4 杯
蠔油 1 湯匙

1/4 cup chicken broth
1 tbsp oyster sauce

荗汁 | Thickening

生粉 1/2 茶匙
水 2 湯匙

1/2 tsp caltrop starch
2 tbsps water

做法 | Method

1. 雞翼洗淨，瀝乾水分（如用雪藏雞翼需要預先解凍），下醃料醃 10 分鐘。
2. 雲耳、金針、冬菇分別浸透，洗淨，瀝乾水分，去蒂。雲耳切小朵。
3. 雞翼放蒸碟上，雲耳、金針、冬菇、杞子加調味料拌勻，放在雞翼上。
4. 在鑊中放蒸架，加水至接近蒸架高度，燒滾水，將雞翼用大火隔水蒸約 8 分鐘至熟即成。

1. Wash chicken wings (defrost if they are frozen) and drain. Marinate for 10 minutes.
2. Soak thoroughly black fungus, dried lily flowers and dried black mushrooms respectively. Wash and drain. Discard stalks. Cut black fungus into small pieces.
3. Put chicken wings onto a dish, add marinade in black fungus, dried lily flowers, dried black mushrooms and wolfberries. Mix well and put on chicken wings.
4. Place steam rack into wok, add water to nearly the height of the rack. Boil water, steam chicken wings over high heat for 8 minutes. Serve.

入廚貼士 | Cooking Tips

- 雞翼加薑汁酒醃，可去除雪藏味。
- Marinate chicken wings with ginger juice to remove the frozen taste.

電飯煲鹽焗雞

Salt-roasted Chicken in Rice Cooker

4~6 人
Serves 4~6

30 分鐘
30 minutes

材料 | Ingredients

光雞 1 隻	1 chicken
粗鹽 5 湯匙	5 tbsps coarse salt
幼鹽 1 湯匙	1 tbsp fine salt
薑 3 片	3 ginger slices
蔥 1 棵	1 sprig spring onion
胡椒粉少許	some pepper

做法 | Method

1. 光雞洗淨，瀝乾水分，用幼鹽和胡椒粉擦遍全身內外。
2. 薑、蔥洗淨，塞入雞腔內。
3. 用錫紙包住電飯煲內鍋，將粗鹽均勻地鋪在鍋底，將雞平放在粗鹽上，按掣焗至熄火。
4. 將雞身轉向另一面，再按掣焗至熄火，再煮 5 分鐘。當水滴落時，電飯煲會自動亮燈，焗至熄火，取出，稍攤涼後斬件上碟。

1. Wash chicken and drain. Smear fine salt and pepper inside and outside.
2. Wash and stuff ginger and spring onion inside.
3. Wrap inside of the cooker with aluminum foil, dash coarse salt evenly at the bottom, place chicken onto coarse salt, press button to cook. Roast till off.
4. Turn chicken to the other side, press button to cook again, roast till off, cook for another 5 minutes. When water dips, the light of rice cooker will be on automatically, roast till off. Take out, cool down slightly. Chop and serve.

入廚貼士 | Cooking Tips

- 雞隻可用食用砂紙包裹，避免鹽附於雞身。
- Chicken can be wrapped in sandpaper so that there'll be no salt on the chicken.

4~6 人
Serves 4~6

90 分鐘
90 minutes

醉雞卷
Drunken Chicken Roll

材料 | Ingredients

雞腿（去骨）1 隻　　1 chicken leg (boneless)

湯料 | Marinade

枸杞 10 粒	10 wolfberries
淮山 5 片	5 Chinese yam slices
圓肉 5 粒	5 dried longans
人參鬚少許	some ginseng
清水 3 杯	3 cups water

調味料 | Seasonings

米酒 20 毫升	20 ml rice wine
紹興酒 20 毫升	20 ml Shaoxing wine
玫瑰露酒 20 毫升	20 ml rose wine
鹽 10 克	10g salt
雞粉 2 茶匙	2 tsps chicken powder

做法 | Method

1. 雞腿洗淨去骨，用錫紙包捲成圓條形，在兩頭紮緊，放入鍋中隔水蒸 20 分鐘，取出，放涼，撕去錫紙。
2. 湯料放入鍋中煮滾，轉慢火再煮 25 分鐘，熄火。
3. 待湯料冷卻，再放入調味料拌勻，將蒸好的雞腿浸約 1 天，即可取出，切片。

1. Wash chicken leg and remove bones. Wrap into cylinder shape with aluminum foil and tie at both ends. Steam in a pot for 20 minutes. Take out and leave to cool, remove aluminum foil.
2. Boil soup ingredients in a pot, and boil over slow heat for another 25 minutes, remove from heat.
3. Wait until the soup is cool, add seasonings and mix. Soak steamed chicken leg for about 1 day, take out and slice.

入廚貼士 | Cooking Tips

- 可以雞扒代替雞腿，省卻去骨的步驟。
- Chicken leg can be substituted by chicken fillet to save the process of removing bones.

豉汁蒸排骨

Steamed Spareribs in Black Bean Sauce

材料 | Ingredients

排骨 300 克
豆豉 1 湯匙

300g spareribs
1 tbsp black beans

4~6 人
Serves 4~6

30 分鐘
30 minutes

◯◯ 調味料 | Seasonings

蒜茸 2 茶匙	2 tsps minced garlic
薑茸 1 茶匙	1 tsp minced ginger
生抽 1 茶匙	1 tsp light soy sauce
粟粉 1 茶匙	1 tsp cornstarch
糖 1/2 茶匙	1/2 tsp sugar
油 1/2 茶匙	1/2 tsp oil
胡椒粉少許	some pepper

◯◯ 做法 | Method

1. 豆豉洗淨，瀝乾水分，剁碎。
2. 排骨洗淨，瀝乾水分。
3. 將豆豉碎加入排骨中，再下醃料醃 25 分鐘，將排骨排在蒸碟上。
4. 在鑊中放蒸架，加水至接近蒸架高度，燒滾水，將排骨用大火隔水蒸約 15 分鐘即成。

1. Wash black beans and drain, chop.
2. Wash spareribs and drain.
3. Add mashed black beans into spareribs, marinate for 25 minutes. Put spareribs onto a dish.
4. Place steam rack into wok, add water to nearly the height of the rack. Boil water, steam spareribs over high heat for about 15 minutes. Serve.

入廚貼士 | Cooking Tips

- 清洗排骨時，放在水龍頭下沖洗，可令肉質較爽。
- Rinse spareribs under running tap to make it crunchier.

4~6 人
Serves 4~6

25 分鐘
25 minutes

Steamed Beancurd with Minced Pork

肉碎蒸豆腐

材料 | Ingredients

絞豬肉 150 克	150g minced pork
豆腐 1 磚	1 piece beancurd
葱花 1 茶匙	1 tsp shredded spring onion
油 2 湯匙	2 tbsps oil

醃料 | Marinade

老抽 1 茶匙	1 tsp dark soy sauce
鹽 1/2 茶匙	1/2 tsp salt
糖 1/2 茶匙	1/2 tsp sugar
生粉 1/2 茶匙	1/2 tsp cornstarch

做法 | Method

1. 絞豬肉加醃料醃 20 分鐘。
2. 豆腐洗淨,瀝乾水分,切小塊。
3. 豆腐排放在蒸碟上,鋪上絞豬肉。
4. 在鑊中放蒸架,加水至接近蒸架高度,燒滾水,將肉碎豆腐用大火隔水蒸約 10 分鐘,灑上葱花。
5. 另燒熱 2 湯匙油,熄火,淋在肉碎豆腐上即成。

1. Marinate minced pork for 20 minutes.
2. Wash beancurd and drain, cut into small pieces.
3. Put beancurd onto a dish, place minced pork on top firmly.
4. Place steam rack into wok, add water to nearly the height of the rack. Boil water, steam minced pork and beancurd over high heat for about 10 minutes. Add shredded spring onion.
5. Heat 2 tbsps of oil, remove from heat. Pour over minced pork and beancurd. Serve.

入廚貼士 | Cooking Tips

- 若用盒裝豆腐,要選用蒸豆腐。
- Choose beancud for steaming if it is boxed beancurd.

鹹魚蒸肉丸

Steamed Meat Ball with Salted Fish

◯◯◯ 材料 | Ingredients

梅頭豬肉 250 克	250g tenderloin pork
鹹魚 1 小塊	1 small piece salted fish
馬蹄 4 粒	4 water chestnuts
薑茸 1 茶匙	1 tsp minced ginger

◯◯◯ 醃料 | Marinade

雞蛋白 1 隻	1 egg white
油 1 茶匙	1 tsp oil
生粉 1 茶匙	1 tsp caltrop starch
糖 1/3 茶匙	1/3 tsp sugar
鹽 1/3 茶匙	1/3 tsp salt
胡椒粉 1/3 茶匙	1/3 pepper
水 2 茶匙	2 tsps water

4~6 人
Serves 4~6

25 分鐘
25 minutes

做法 | Method

1. 梅頭豬肉洗淨，瀝乾水分，剁碎後放大碗中，加醃料攪拌至起膠。
2. 馬蹄去皮，洗淨，瀝乾水分，切粒。
3. 鹹魚洗淨，瀝乾水分。
4. 燒熱油鑊，下鹹魚煎至金黃，盛起，待涼後起肉，切碎。
5. 將所有材料放大碗中拌勻，順一個方向拌至起膠，用虎口擠出肉丸，排放在蒸碟上。
6. 在鑊中放蒸架，加水至接近蒸架高度，燒滾水，將肉丸用大火隔水蒸約 15 分鐘即可。

1. Wash tenderloin pork and drain. Put into a big bowl after chopped, add marinade and stir till elastic.
2. Peel water chestnuts and wash. Drain and cut into dice.
3. Wash salted fish and drain.
4. Heat oil in wok, shallow-fry salted fish till browned. Take out, remove bones after cooling down and then strip.
5. Put all ingredients in big bowl and blend in one direction till elastic. Squeeze out meat balls with hands, put onto the dish.
6. Place steam rack into wok, add water to nearly the height of the rack. Boil water, steam meat balls over high heat for about 15 minutes. Serve.

入廚貼士 | Cooking Tips

- 豬肉攪拌後再用手撻向碗中，更易起膠。
- Smash minced meat into the bowl with hands to make it elastic.

釀冬菇

Stuffed Dried Mushrooms

⦾⦾⦾ 材料 | Ingredients

花菇 8 朵	8 dried black mushrooms
梅頭豬肉 80 克	80g tenderloin pork

⦾⦾⦾ 豬肉醃料 | Marinade for Pork

生抽 1 茶匙	1 tsp light soy sauce
糖 1/2 茶匙	1/2 tsp sugar
粟粉 1/2 茶匙	1/2 tsp cornstarch

⦾⦾⦾ 冬菇醃料 | Marinade for Black Mushroom

生粉 1 茶匙	1 tsp caltrop starch
油 1 茶匙	1 tsp oil
糖 1/2 茶匙	1/2 tsp sugar

芡汁料 | Thickening

清雞湯 3 湯匙	3 tbsps chicken broth
蠔油 1 湯匙	1 tbsp oyster oil
粟粉 2 茶匙	2 tsps cornstarch
麻油 1 茶匙	1 tsp sesame oil

做法 | Method

1. 冬菇浸軟，去蒂，洗淨，瀝乾水分，加冬菇醃料略醃。

2. 梅頭豬肉洗淨，瀝乾水分，剁碎後放大碗中，加豬肉醃料順一個方向攪拌至起膠。

3. 將梅頭豬肉碎釀在冬菇中，排放在蒸碟上。

4. 在鑊中放蒸架，加水至接近蒸架高度，燒滾水，將釀冬菇用大火隔水蒸約 15 分鐘，熄火。

5. 芡汁料放碗中拌勻。另燒熱油鑊，下芡汁料煮滾，淋在釀冬菇上即成。

1. Soak dried black mushrooms till soft, remove stalks. Wash and drain. Add dried black mushroom marinade and mix well.

2. Wash tenderloin pork and drain. Put into a big bowl after chopped, add pork marinade and stir in one direction till elastic.

3. Stuff minced tenderloin pork into dried black mushrooms, then place onto a dish.

4. Place steam rack into wok, add water to nearly the height of the rack. Boil water, steam stuffed shiitake mushrooms over high heat for about 15 minutes. Turn off heat.

5. Mix thickenings in bowl. Heat oil in wok, pour in thickening and cook. Pour onto stuffed dried black mushrooms and serve.

入廚貼士 | Cooking Tips

- 免治豬肉以細切大剁的方式剁成，較用攪拌機攪出的有口感。
- The texture of minced pork would be better if it's chopped from slices compared with those minced by machine.

梅菜蒸肉片

Steamed Pork Slice with Preserved Cabbage

◯◯◯ 材料 | Ingredients

瘦梅頭豬肉 300 克
甜梅菜 1 棵

300g lean tenderloin pork
1 stalk sweet preserved cabbage

◯◯◯ 豬肉醃料 | Marinade for Pork

粟粉 1/2 茶匙

1/2 tsp cornstarch

4~6 人
Serves 4~6

15 分鐘
15 minutes

梅菜醃料 | Marinade for Preserved Cabbage

糖 1 茶匙	1 tsp sugar
油少許	some oil

做法 | Method

1. 梅頭豬肉洗淨，瀝乾水分，切片，加豬肉醃料拌勻。

2. 甜梅菜用水浸片刻，洗淨，瀝乾水分，切粒。

3. 燒熱鑊，以白鑊略炒梅菜粒，盛起，加梅菜醃料拌勻。

4. 梅頭豬肉與梅菜拌勻，平放蒸碟上。

5. 在鑊中放蒸架，加水至接近蒸架高度，燒滾水，將豬肉用大火隔水蒸約 7 分鐘即成。

1. Wash lean tenderloin pork and drain. Slice, add pork marinade and mix well.

2. Soak sweet preserved cabbage for a while. Wash and drain, cut into slices.

3. Heat wok, stir-fry preserved cabbage dices slightly, take out and add preserved cabbage marinade, mix well.

4. Blend tenderloin pork and preserved cabbage, place it onto a dish flat.

5. Place steam rack into wok, add water to nearly the height of the rack. Boil water, steam pork over high heat for about 7 minutes. Serve.

入廚貼士 | Cooking Tips

- 梅菜中可能有沙粒，必須撕開菜夾洗淨。
- There maybe sands in preserved cabbage, tear it open and wash.

4~6 人
Serves 4~6

18 分鐘
18 minutes

土魷蒸肉餅

Steamed Minced Pork with
Dried Squid

材料 | Ingredients

絞豬肉 300 克
土魷 3 隻
冬菇 3 朵
馬蹄 3 粒

300g minced pork
3 dried squids
3 dried black mushrooms
3 water chestnuts

醃料 | Marinade

生抽 2 茶匙	2 tsps light soy sauce
粟粉 2 茶匙	2 tsps cornstarch
糖 1/2 茶匙	1/2 tsp sugar

做法 | Method

1. 土魷浸軟，撕去軟骨，瀝乾水分，切粒。
2. 冬菇浸軟，洗淨，去蒂，瀝乾水分，切粒。馬蹄去皮洗淨，瀝乾水分，切粒。
3. 絞豬肉放大碗中，加醃料拌勻，順一個方向打至起膠，再加入其他材料拌勻，平鋪在蒸碟上。
4. 在鑊中放蒸架，加水至接近蒸架高度，燒滾水，將肉餅用大火隔水蒸約 8 分鐘即成。

1. Soak dried squids till soft, tear off the soft bones. Drain and dice.
2. Soak dried black mushrooms till soft, wash and remove stalks. Drain and dice. Peel water chestnuts and wash. Drain and dice.
3. Place minced pork in a big bowl, add marinade and mix in one direction till elastic. And then add other ingredients and mix, put it onto a dish flatly.
4. Place steam rack into wok, add water to nearly the height of the rack. Boil water, steam pork over high heat for about 8 minutes. Serve.

入廚貼士 | Cooking Tips

- 攪好的肉餅可加 1/2 隻蛋白，使肉餅較滑和不會收縮。
- Add 1/2 egg white into the minced pork can make it smoother and avoid shrinking easily.

榨菜蒸牛肉

Steamed Beef with Preserved Mustard Head

材料 | Ingredients

牛肉 300 克
榨菜 1 小塊

300g beef
1 small piece preserved mustard head

牛肉醃料 | Marinade for Beef

生抽 2 茶匙
粟粉 1 茶匙
糖 1/2 茶匙

2 tsps light soy sauce
1 tsp cornstarch
1/2 tsp sugar

2~4 人
Serves 2~4

20 分鐘
20 minutes

榨菜醃料 | Marinade for Preserved Mustard Head

糖 1 茶匙
麻油 1 茶匙

1 tsp sugar
1 tsp sesame oil

做法 | Method

1. 榨菜洗淨，瀝乾水分，切薄片，加榨菜醃料拌勻。
2. 牛肉洗淨，瀝乾水分，切片，加牛肉醃料醃 5 分鐘，加 1 湯匙水拌勻，再醃 5 分鐘。
3. 榨菜和牛肉相間排在蒸碟上。
4. 在鑊中放蒸架，加水至接近蒸架高度，燒滾水，將牛肉用大火隔水蒸約 5 分鐘即成。

1. Wash preserved mustard head and drain, cut into thin slices, add preserved mustard head marinade and mix well.
2. Wash beef and drain, slice. Add beef marinade and wait for 5 minutes. Add 1 tbsp of water and blend, marinate for another 5 minutes.
3. Place preserved mustard head and beef alternately onto a dish.
4. Place steam rack into wok, add water to nearly the height of the rack. Boil water, steam beef over high heat for about 5 minutes. Serve.

入廚貼士 | Cooking Tips

- 醃牛肉時加入水分，可令牛肉鬆化、幼滑。
- Add water when marinating beef to make it smooth and soft.

蒸出美味

編著
梁綺玲

編輯
紫彤

美術設計
Nora

排版
何秋雲

翻譯
Mo Yu Chun

攝影
Fanny

出版者
萬里機構出版有限公司
香港鰂魚涌英皇道1065號東達中心1305室
電話：2564 7511
傳真：2565 5539
電郵：info@wanlibk.com
網址：http://www.wanlibk.com
　　　http://www.facebook.com/wanlibk

發行者
香港聯合書刊物流有限公司
香港新界大埔汀麗路36號
中華商務印刷大廈3字樓
電話：2150 2100
傳真：2407 3062
電郵：info@suplogistics.com.hk

承印者
美雅印刷製本有限公司

出版日期
二零一八年五月第一次印刷

萬里機構

萬里 Facebook